Birds

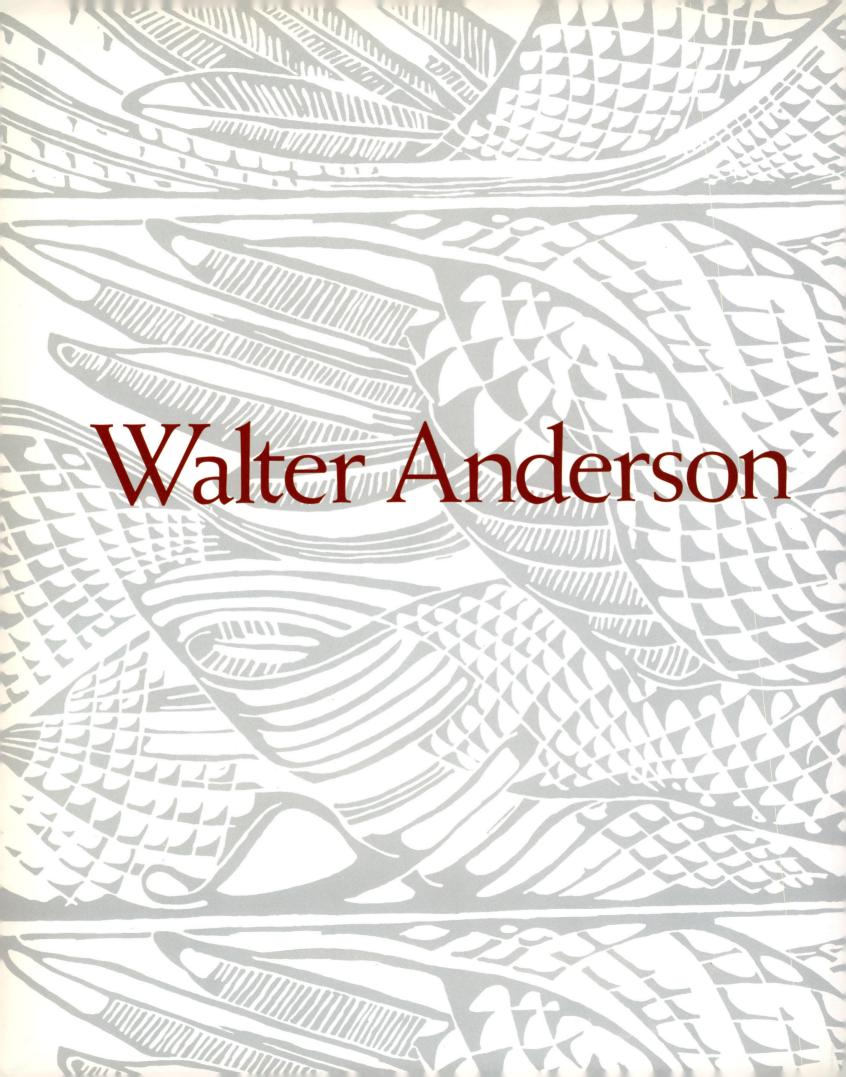

Walter Anderson

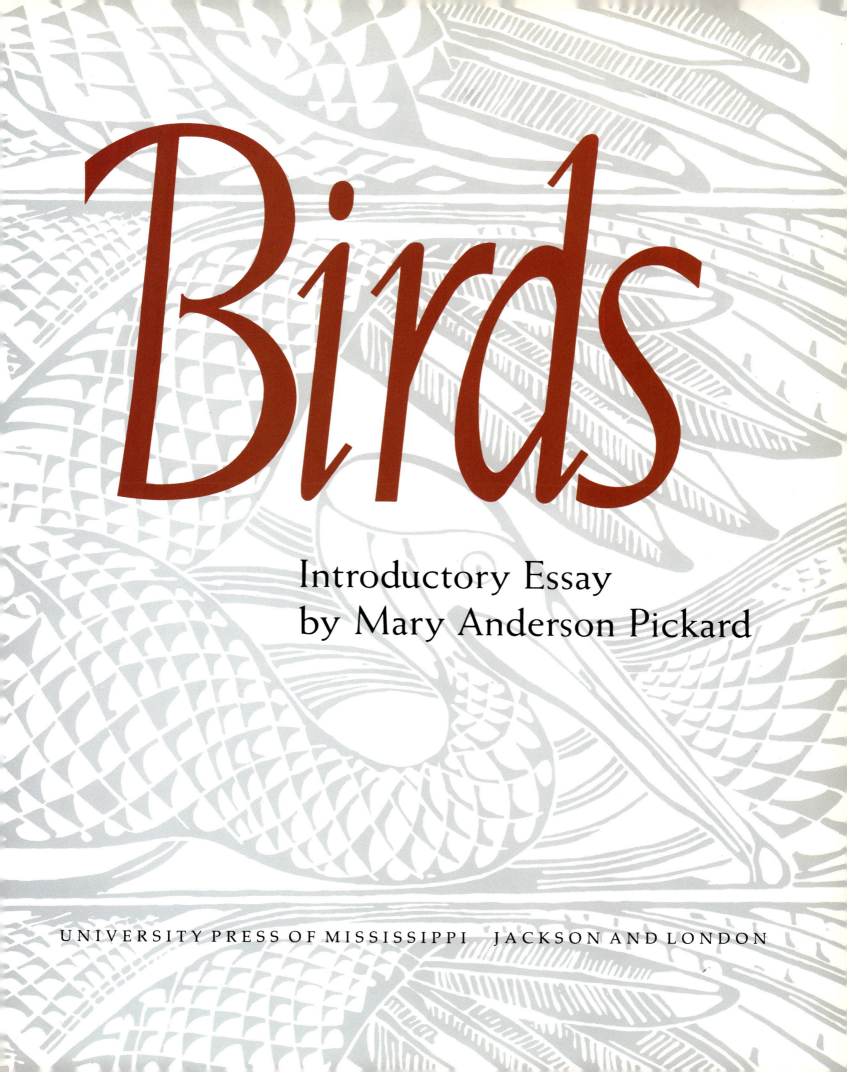

Birds

Introductory Essay
by Mary Anderson Pickard

UNIVERSITY PRESS OF MISSISSIPPI   JACKSON AND LONDON

Publication of this book has been assisted by the generosity of the *Director's Circle of the University Press of Mississippi.*

Alcorn State University
Bookfriends of the University Press of Mississippi
Delta State University
Deposit Guaranty National Bank
Dockery Farms
Eastover Bank for Savings
Jackson State University
John and Harriet DeCell Kuykendall
LeMoyne College
McCarty Farms
Mississippi Power Company
Mississippi Power and Light Company
Mississippi State University
Mississippi University for Women
Mississippi Valley State University
Mississippi Valley Title Insurance
Mobile Communications Corporation
Mobile Telecommunications
Pepsi Cola Company of Jackson
Phil Hardin Foundation
South Central Bell
Sunburst Bank
University of Mississippi
University of Southern Mississippi
W. E. Walker Foundation

Designed by John A. Langston

British Library Cataloguing in Publication data available

The text on the section opening pages comes from the writings of Walter Anderson.

CIP data on page 118

*This book is dedicated
to those who find
in Walter Anderson's words and images
stimulus for their own
transcendent flights*

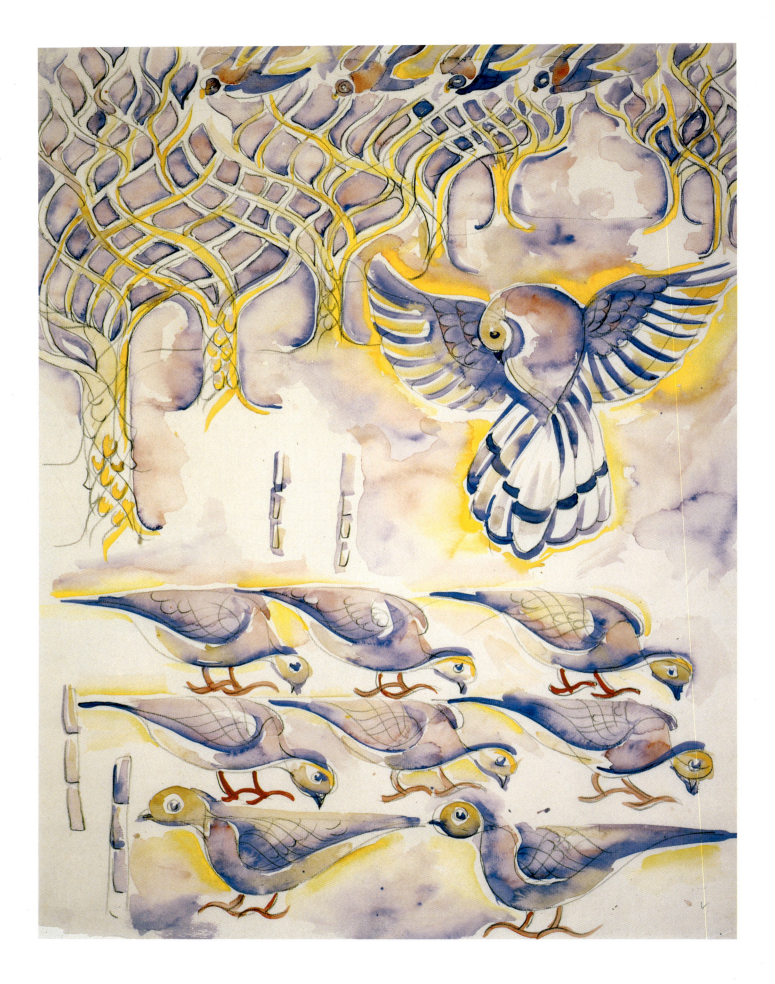

The Birds of Walter Anderson

by Mary Anderson Pickard

Birds quicken my earliest memories of my father, Walter Anderson.

A kingfisher nested unexpectedly in the hollow limb of a great hurricane-felled live oak that sprawled on our beach at Oldfields, in Gautier, Mississippi. My father, returning from an early walk, heard its cry and saw it leave the hole. Climbing a lower branch, he discovered its secret. He came home exalted and, eager to share the miracle, hoisted his four-year-old daughter to his shoulders and strode back to the big log. I clutched his neck with my knees and grabbed at a handful of shaggy brown hair. "Gently," he cautioned as we climbed the tree, and again, "gently," as I lurched forward to peer into the long dark opening at five perfect white eggs. There was a racking shriek, and the kingfisher attacked, a blur of gray and white, a flash of rose! We leapt, landed in the sand, and galloped off in awkward retreat. It was more than we could tell the others at breakfast.

Chickens, ducks, geese, turkeys and guinea hens escaped from the barnyard every day and invaded the green lawn around our house at Oldfields. They pecked, waddled, ran, postured and strutted. My father, sitting cross-legged on the ground, his clipboard resting on his knee, watched them with delight. His crow-quill pen dipped into the bottle of ink, and I marveled at the sudden appearance of feathers, beak, claws and beady eye. Soon, however, I wandered away, tired of the repetition and his exclusive concentration. He had forgotten I was there. I accepted his drawing, not as his work or play, but as his endless preoccupation.

Gulls, terns and pelicans claimed the long pier in front of Oldfields as well as the sky above the green-gold waters of the Pascagoula Sound. They perched on the old posts and whitened the silvered wharf planks; they wheeled and cried against the clouds, the water, and the long low scallop of Horn Island's trees on the horizon. How they soared in the updraft above the bluff in a white summer squall. Elated, my brother Billy and I hurled ourselves against the wind and ran, arms spread or flapping in awkward imitation. In the purple light at the end of the pier, Daddy drew them. I could tell from his drawings that he knew how they flew.

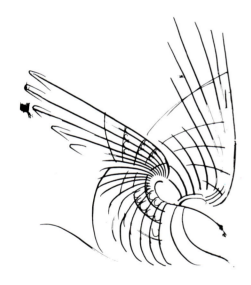

Flight c. 1940

Wind Wave Bird C. 1945

Sandpipers ran down the beach ahead of Billy and me when we accompanied our father on arrowheading expeditions to the eroded Indian mounds at Graveline. Unlike our mother, he never shortened his step for our benefit so he was usually far ahead of the sandpipers. By the time we trudged around the point he'd be out of sight. Were we deserted? But a bird-like call from the woods above the beach had us scrambling up the bluff to see his latest discovery or—more likely—to play his unique version of hide-and-seek. He always hid, then signaled his whereabouts with bird sounds. We tried hard to walk noiselessly like Indians in our clumsy search, but when we reached the certain tree or bush where we were sure he was hiding, he had inevitably flitted on and the next peep or squawk sent us off in the opposite direction. Only when we were exhausted and crying from rockachaws in our feet would he reveal himself, delighted with having fooled us. Other times he'd lose interest and take a nap or forget our presence and start home without us. Finally we'd realize he was gone and trail after, dragging our feet and sniffling. Usually we discovered him a mile down the beach sitting against a log drawing sandpipers.

Red-winged blackbirds sang in the reeds that edged the bayou where I sat with Billy and my father neck deep in brackish water. "Gently," said Daddy and "quietly." Gradually on our adventures with him we learned to be still, to wait silently for the moment of revelation in the natural world. Glimpses of the feeding green heron's serpentine strike, the ghost-like rail with her fluffy chicks, glorious wood ducks and comical coots were our rewards for patience.

Once my father took me to Fountain Bleau beach to camp for the night. We took sandwiches for our supper and made beds in the sand by digging depressions and lining them with blankets. It was Sunday and at home we had listened to the radio, as we always did, to the New York Philharmonic and Daddy's beloved Beethoven. In the evening we sat at the water's edge to eat our sandwiches. Terns came for the crumbs we tossed, diving and rising to the waves' movement and the wind's lift. My father spoke of music, of the various instruments, the musicians, how the parts fit together to form a symphony. I felt the damp wind on my face and threw my last crust to the nearest tern.

"Listen," he said suddenly.

I listened. The rise and fall of the wind, the plash of the waves, the punctuating cries of the gulls were all parts of a whole. My father stood and raised his arm as though he were conducting. The curves of the evening clouds, the straight long line of the horizon, Horn Island's syncopated pattern of tree trunks and spaces, the diagonal wavy lines of water, the spiral curls of the wave tops, the bubbles on the wet scalloped sand all fit together and were balanced and emphasized by my father, dark and vertical, his feet on the sand, his head against the bright sky as the sun went down.

When I was a child my mother, Agnes Anderson, was my very adequate parent. My father, when he was there, kept things off balance and exciting, sometimes frightening, but never dull. From 1940 until 1946 at Oldfields he was a concentrated presence in my life whose intense enthusiasms for art, literature, music and nature were illuminating and contagious. Although he worked harder than anyone I have ever known at building a rental cottage or a pier, gardening and getting wood, his energy was focused on his art. His tensions, his unpredictable comings and goings foiled my mother's efforts to serve meals at regular hours and establish a daily routine for her family. His day invariably included drawing. Otherwise his activities were consistent only with the nature of the day and his changeable mood.

From 1947 until his death in 1965 he lived apart from us in his cottage at Shearwater Pottery in Ocean Springs or on Horn Island. My mother, my sister Leif, my brothers Billy and John and I lived nearby with his mother in the renovated barn that was her home. We saw him only at his inclination and that wasn't often. On those increasingly rare occasions when our comings and goings did coincide, I was expected to treat him with respectful affection. "Give your father a kiss, Mary," my grandmother would say when I burst in from school on one of his infrequent visits to her at the barn. I think it was as awkward for him as it was for me. I remember his careless dress, his ancient and strangely colored shirts and trousers, his broken sometimes mismatched shoes on sockless feet, the inimitable felt hat shaped by weather and use.

From 1954 until I finished college I saw him only on holidays when he was very interested in books I was reading for literature courses. At summer's end I retrieved from his cottage my books he'd borrowed, or I never saw them again. After 1959, marriage and family and teaching demanded my full attention, and then, in 1965, he died. I had hardly known him all those years. With the rest of my family I was astonished and moved by the treasure we found in his dilapidated cottage. In 1968 I viewed with overwhelmed delight his first retrospective exhibition, "The World of Walter Anderson," at Brooks Museum of Art in Memphis. Gradually the public interest and response grew. In 1975 my mother, wearying of the growing burden, asked me if I would like the job of caring for the collection. At first I continued to teach during the day while making an inventory of the watercolors at night as my own children slept. I began with the birds.

The watercolors had been named and numbered by my mother, her sister, Pat, and my aunt Sara, who was the real bird watcher in our family. They had given some order to the chaotic piles and boxes of typing paper, but nothing was sorted. Ducks, sparrows, pelicans—sketches and finished pieces—mingled in the vast dissimulation of birds. No rag paper separated and protected them. No acid-free folders or boxes housed the ordinary manila folders that were stuffed with watercolors.

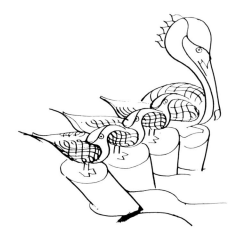

Pelican and Terns   c. 1940

I opened a folder. Perched on the branch of a dead pine on Horn Island, a kingfisher surveyed his domain. My father had used his brush to print "The Halcyon Bird." I remembered the familiar myth he'd told us and his habit of calling a shining winter day "halcyon." That first evening I couldn't stop looking.

Brown pelicans filled the skies over Chandeleur Island, "rising," as he'd written in his journal, "in tremendous musical harmonies." A dead lesser scaup, washed ashore on the Horn Island beach, glowed iridescent in a careful still life.

Two mated green herons attended their marshy nest under a yin/yang sun reiterating their union.

Red-winged blackbirds gobbled rice. Hummingbirds jeweled the fullness of blooming thistles. Migrating redstarts rested among mangrove tubers.

After a while I no longer saw birds, only color and movement. My eyes

Childhood Drawing   c. 1910

were too full, and my head ached dizzily. In bed I was unable to sleep. In the pictures of dabbling ducks with their feet up and heads down but visible through the clear water of the Horn Island Lagoon, I had recognized an appreciative sense of humor. In the careful studies of fledgling birds, nakedly exposed, I sensed the ruthless, probing intelligence I remembered, his curiosity and his concentrated focus. Through the birds I was rediscovering my father.

Drawings from his childhood packed away in an old footlocker show my father's early interest in birds. Looking at them, I can hear his mother's words to me: "If you want to draw a bird, start with an egg."

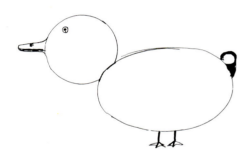

Egg Duck   c. 1908

With a firm stroke of her pencil she'd place an egg on my paper which I was to augment with head, wings and tail. Similarly, in a battered tablet young Walter had transformed eggs and a variety of other shapes—including a hairbrush he had traced—into a convincing if unlikely assortment of ducks and song birds. Years later, he wrote on the back of a drawing, "The outpouring of life must find expression in the egg."

Annette McConnell Anderson, my father's mother, believed in the discipline of drawing every day. Her own considerable talents for music, writing and art were never fully developed. Recognizing her own lack, she tried to instill good work habits in her sons, and later in her grandchildren, by presenting them with sketchbooks in which they were to draw a flower, a leaf, or a bird each day. She believed that writing skills could be developed in the same way. I found among my father's sketchbooks one of her notebooks marked distinctively "200 words a day." Crossing out "words" he'd written "birds" and filled the pages with sketches of egrets, osprey and sandpipers.

The Anderson home in New Orleans, where my father grew up, was located at 553 Broadway. The duck ponds and live oaks of nearby Audubon Park provided birds enough for many notebooks. With their father, Walter and his brothers Peter and Mac hunted and fished on Lake Pontchartrain and in the vast marshes of Louisiana on weekends. Like many naturalists, Walter learned to recognize water fowl over the sights of a shotgun.

Birds on the Beach   C. 1934
(verse by Agnes Anderson, miscopied by artist)

Summers meant the shore birds of the Mississippi Gulf Coast at Bay St. Louis where Annette's father had a house on the beach. There the boys learned to swim and sail and love the offshore islands.

In 1915, twelve-year-old Walter, a homesick student at the Manlius Military School in New York state, wrote the required brief letters home. Those notes reveal a sustaining and comforting awareness of nature and an interest in birds:

> I go for a walk in the woods nearly every day and I have special places where I go and read and draw.

> It was ever so good of you to send me those books. . . . I am going out drawing this evening with the book that you sent me.

> Yesterday I went for a walk in the woods and I found a partridge's nest with twelve large cream colored eggs in it. Bradley is going to draw a picture of it for me.

> I think I know where a pheasant's nest is, and a boy named Willard is going to take a photograph of it. He also took a picture of the partridge's nest with the eggs in it . . . yesterday . . . I found two robins' nests and a cat bird's nest.

In 1935 my father's interest in birds was stimulated by a new acquaintance whom he met while visiting in Baltimore with his young wife, Agnes. Dr. Ned Park was an enthusiastic birder who presented Walter with his first guide to birds, *Chapman's Color Key to North American Birds*. Watching birds together, both men deplored the lack of a guide which dealt specifically with birds of the Southeast. Encouraged by Dr. Park, Walter began work on a series of plates. He wanted them to be ornithologically correct and decided that, like other bird illustrators, he would work from dead specimens. He asked his younger brother Mac to shoot them for him. My mother remembers that the killing of the birds for the ornithological illustrations bothered him from the beginning. Increasingly he regretted his responsibility for their deaths. He made careful realistic drawings and colored them in sensitive watercolor wash,

Warbler   C. 1935

but the work coincided with a period of deepening depression. The strain of his regular job at Shearwater Pottery, which left little time for drawing and painting, contributed to a growing tension and severe headaches. A long bout with undulant fever in the late fall of 1935 followed by the death of his father in February 1936 proved devastating. Periods of complete withdrawal and an attempted suicide preceded his mental breakdown in 1937.

Until the breakdown his interest in birds had been intellectual and objective. Naturally curious and fascinated by the infinite variety of form and color in nature, he studied birds as he did his collections of moths and butterflies and the indigenous plants he located in the surrounding woods and transplanted to the grounds at Shearwater.

He spent most of the next three years, from 1937 until 1940, in mental hospitals. In the world's eyes he'd failed as a husband and provider—as a member of society; in his own eyes, he'd failed his art. His loss of freedom and the severe drug-induced shock treatments he was given augmented the anguish of his illness. From the ordeal he would emerge with a new and lasting respect and reverence for life and liberty.

Wild Ducks over Philadelphia   c. 1939

In the hospital he withdrew for months into complete apathy. He gradually began to draw again during analysis. Many of the drawings are of incidents in his past but seen from a bird's perspective. In identifying with birds he could rise above the hospital and the world's demands and judgments and find a needed respite and objective point of view. In one drawing, as a heron, he soars above the familiar maze of bayous in a Louisiana marsh; in another, as a gull, he views old Biloxi schooners racing down the bay. Wild ducks frame a statue of William Penn as they survey the Philadelphia he knew from his years at the Pennsylvania Academy of the Fine Arts. With a pair of widgeons over the marshy prairie he looks down on the hunters of his youth.

Early in 1939 my father abruptly left the Shepherd Pratt Hospital in Towson, Maryland, and walked home to Ocean Springs. No one knew where he was for months. When he finally arrived at Shearwater, he was unable to tolerate living with my mother and me. He spent most of his time on a platform that he built in a tall pine tree across the road from the cottage. He called it his aerie. (Among the hospital drawings is a series showing a blue jay family nesting and raising their young. Below in the yard of the cottage he and his own family are visible through the branches as they come and go about their lives.)

Confined briefly in the state hospital at Whitfield, he refused to stay. According to an apocryphal story, he escaped from an upper window leaving soap-drawn birds flying on the dark brick wall. Living in Jackson while being treated as an outpatient, he wrote illustrated letters to Sissy in which ducks, pelicans, or mockingbirds look down on an idyllic life: Bob and Sissy living alone—without babies—at the cottage.

The mental hospital drawings were discovered in 1985 in an old suitcase stored at his brother Peter's house. In addition to the bird's-eye views of my father's illness and his past life, the pictures reveal his awareness and understanding of birds as powerful and ambivalent symbols in literature and in world mythology.

Fragile and vulnerable, my father finally rejoined us at Oldfields in 1940. He found his senses heightened to an extraordinary awareness. On a fragment of paper from the Oldfields years he wrote, "Like this year's bird born in

Walter and Agnes at the Cottage C. 1939

December 31 Calendar Drawing c. 1942

Vireo c. 1940

the mountains, I had never seen the sea before." Fired by his new perceptions he worked urgently "to understand significant form" in drawings. He wrote:

> I am continually arriving from some strange place and everything I see is new and strange. . . . The realization of form and space is through feeling. When I feel the beauty of a flower or the trunk of a tree I am at once inducted into a world of three dimensions and have a sense of form which is the opposite of artificial forms and conventions. I live and have my being in a world of color and shape. Consciousness of this means being alive. You arrive at that consciousness through the five senses: smell, sight, hearing, touch and taste.

He celebrated each day in calendar drawings. My mother's birthday, the planting of the garden, the first robin or violet or a new calf were subjects for the daily drawings done with the regularity of his childhood work in notebooks. Symmetrical and delicately colored in flat wash, the pictures mirror the relative peace of those few years. Birds occur frequently in the appropriate seasons and indicate the setting of the artist's activities on a particular day— the woods, the barnyard, the beach, an expedition up the river or to an island. His revived interest in art history resulted in stylistic variety in the calendar drawings. An Oldfields egret looks quite Egyptian. A Babylonian-like frieze features ducks, turkeys and chickens from the barnyard. Shore birds fill their rectangle with the compacted complexity of a Mayan stele.

A Method for Creative Design, written by art historian Adolpho Best-Maugard, reduced all art to seven organic motifs: the spiral, the circle, the half circle, the S-curve, the wavy line, the zig-zag and the straight line. My father adopted the seven motifs and carried them to incredible lengths in studies of feathers and flight. They are evident in the strong patterns of the large linoleum blocks depicting Gulf Coast flora and fauna, which he produced for the public.

Walter on Oldfields Pier   C. 1940

The Gulf Coast has a large bird population. Its proximity to the Mississippi flyway means that the regular numbers are swelled twice each year by migratory species. Shorebirds abound and many varieties of wild ducks winter in the estuaries and on the barrier islands. At Oldfields my father used the big Gilbert Pearson *Birds of America* to mark sightings and locations, especially in spring and fall. He kept bird lists scratched on the backs of drawings or on scraps of paper bags or wrapping paper.

Just a few miles north of Oldfields between Gautier and Vancleave, sandhill cranes have nested for many years. (Later the completion of the I-10 highway was actually delayed for more than a year by bird lovers who succeeded in having a large acreage of cutover land set aside as a refuge.) In January 1942 my father made a walking tour (he would have called it a progress) to visit and draw the cranes. As was his habit, he kept a journal of the trip. He left the house at five o'clock, well before daylight. The ground was frosty.

Cranes   C. 1945

> I walked between the trunks of trees and watched the stars appear and disappear as I passed. . . .

> The dawn was magnificent. The sky was vermillion and blue and all the tree trunks turned pink and the grass peach colored. Then I heard the cranes for the first time, metallic and in perfect keeping with the setting . . . something between bells and brass horns. . . . It was a musical progression, recurring intervals of pine ridges, low land, tupelo and cypress, water, then a ridge with scrub pine . . . and an accompaniment of cranes. After crossing one branch I came out into a beautiful stretch of open country . . . looked up and saw one of the most heavenly sights I've ever seen. Walking along, with their heads just below the horizon, were eight enormous cranes. They were strung out, three and then five. I stopped and drew them.

His trip lasted several days. He camped at night and walked and drew. After lunch on the second day he described an experience that illuminates his extraordinary relationship with birds.

> I was on the point of leaving when a strange and secret thing happened. I felt I needed a sign that the birds still loved me and I thought if only one of you will come a little closer I will know, and at once a hermit thrush came and sat on a stump a few feet from me and a woodpecker came and lit on a trunk. . . . Then suddenly I knew the meaning of love . . . that love meant having an object to love and not reception, any amount of reception, could possibly give understanding in the same way.

II

Pelican Lampstand   c. 1930

Birds were an integral part of my father's life and work. Of all his birds, pelicans were most conspicuous, fully realized in wood, clay, linocuts, drawings, watercolors and as decorative motifs on pottery. His study of the pelicans was exhaustive. In the 1940s he camped at the Chandeleur rookeries during the nesting seasons, living in the midst of thousands of pelicans, identifying with them, sharing "in all their reactions and conditions of life." He even attempted to record a pelican "Dictionary of Common Terms." On Chandeleur in the pelican world, he appreciated for the first time the unity of life, the separate yet connected strata, each dependent on the other. Years later in a Horn Island log he would reflect: "It is easy for me to appreciate the earlier and more primitive forms of life like . . . pelicans. They really are simpler and more closely related to function. . . . On an island where the means are limited, the forms of nature are limited by those means, and more nearly related to that unity which is represented by the island." Beautiful and ugly, dignified and comical, graceful and awkward, fragile yet indomitable, the pelican is a union of opposites perhaps reflecting the ambivalence of my father's own nature and for that reason endlessly fascinating to him.

In *Approaching the Magic Hour,* my mother's memoirs of her life with my father, she recalls an incident from 1933, the first year of their marriage. For a day and a night he disappeared. When he finally turned up she learned he'd spent the hours on Marsh Point, a muddy peninsula in Biloxi Bay, in the company of a great flock of migrating white pelicans. "In heaven," he told her.

In his work, the pelican more than any other bird symbolizes the Gulf Coast. Believing that the artist owed good, reasonably priced artwork to the public, my father decorated pottery, carved linoleum blocks for prints, and painted public murals. His subjects were coastal flora and fauna and the history of the area. He felt driven to make people aware of the unique qualities of the coast, thereby giving them pride and the desire to conserve those qualities. A sense of place, he felt, would contribute to their self-esteem. The familiar coast pelican was a perfect theme.

Pelican Christmas Card C. 1935

Two Pelicans on Posts C. 1945

Designing and decorating at Shearwater Pottery in the 1930s, he sculpted three designs for a variety of heights and widths in pelican bookends. These are still cast from molds at Shearwater today, as is a graceful pelican lamp base. He also carved intricate relief decoration featuring pelicans into bowls and vases from which molds were made. To adorn ware that his brother, Peter, threw on the wheel, he used a variety of techniques. On an early red slip-covered bowl called "Harvesting the Sea," a pelican flies over the head of a fisherman pulling his net. The process called sgraffito, which involved scraping away the slip to make the decoration, was effective but very time consuming. Pelicans gracefully soaring, bobbing or plummeting downward in the clumsiest of dives appear and reappear in underglaze decoration on plates, saucers, cups and bowls. Some are incised and then painted on top of white slip; some are painted freehand directly onto the brown clay body. One plate shows a mother pelican feeding her chick from her gullet, the baby's head inside her open beak.

Pelicans inspired more blocks than any other subject when he created the linoleum block prints at Oldfields in 1945 and 1946. In one, an angelic vertical pelican raises his wings in flight against the clouds. In another, the stylized feathers of a long horizontal pelican repeat the ripples of the waves behind him. In a lighter frieze, three fly one after another just above the horizon, in the pattern familiar to any coast dweller. The two pelicans seated on posts look very like the wooden totems he would sculpt a few years later. There was also a soaring pelican as seen from below. I remember his pleasure in the pelican series, the unusual pause to savor his accomplishment. He covered the walls of his bedroom with the rhythmic prints, creating a strange and dizzying perspective. Looking up, I was amazed to see the pelicans soaring diagonally across the ceiling! He had interpreted for all of us those "tremendously musical harmonies" described in the Chandeleur logs.

In 1950 my father asked for and was given permission to paint murals on the walls of the Ocean Springs Community Center. The mural is a hymn to

Pelican with Sailing Ship   C. 1960

Man Rowing with Pelicans   C. 1939

the physical attributes of the place—its plants, weather, insects, animals, and birds—and celebrates its historical significance as the first French colony on the coast. On the south wall, above the Indians advancing to welcome D'Iberville and his crew to the rich new land, pelicans fill the sky and soar above the bountiful bay like humorous and beneficent spirits. Surely no bird was ever used more effectively to represent the excitement and promise of a joyous beginning.

When my father was a child, he knew the pelican from Louisiana's state seal, where it was featured as a symbol of the selfless parent who feeds its young from its own flesh. As a young man fishing on the Gulf Coast, he learned to recognize feeding pelicans as a sure sign of fish below the water's surface and to accept the great flocks as an unchanging and integral part of the bounty of the littoral waters.

In the drawings that he made in the mental hospital in 1938 and 1939, pelicans are always included in memories of sailing or fishing on the coast. At Oldfields when he lived with us, they were a presence each day. From the gallery, the beach, or the pier, he drew them in ink, realizing the characteristic silhouette at rest on a pier post or bobbing on the waves. Using spiraling lines, he explored the gesture of their soaring or diving flight.

In the late 1940s there was a change in his behavior. Did his withdrawal from the family and society prompt the exclusive intensity of his obsession with pelicans on North Key in the Chandeleur Islands in the late 1940s and early 1950s? There, thirty miles out in the gulf on a treeless bar covered with mangroves, he produced an astonishing array of drawings and watercolors. At first they were patterned, musical and stylized, showing a strong use of the

Best-Maugard motifs. Then for the first time the focus and concentration that characterized his Horn Island work became consistent in a series of drawings quite different from anything he'd done before. The Chandeleur logs and an account he wrote summing up his experiences of the rookery reveal insights and understanding beyond his previous observations.

On North Key, isolated from the world, he learned to know and identify with pelicans.

Pelican c. 1950

> After you have lived on the island for a while, there comes a time when you realize that the pelican holds everything for you. It has the song of the thrush, the form and understanding of man, the tenderness and gentleness of the dove, the mystery and dynamic quality of the night jar, and the potential qualities of all life.
>
> In a word you lose your heart to it. It becomes your child, and the hope and future of the world depend upon it.

Ironically, it was during these same years that the great flocks of pelicans began to disappear. My father witnessed the decimation of the Chandeleur rookeries. He noted the decreasing number of birds and the eggs that never hatched. He even speculated correctly on the cause—an effluence of DDT from the mouth of the Mississippi River. Before 1950 there were pelicans everywhere on the Gulf Coast. By 1960 a pelican was a rare sight.

Shortly before my father died, the pelicans had begun to return. A note on the back of a watercolor exults, "17 in one flock! July 1965." He died in November.

III

From 1950 until his death in 1965, I saw very little of my father. He camped for weeks at a time on Horn Island, spending far more time with birds than with people. He wrote in his log: "This morning I painted the bittern's nest while the flies stung. Later did a water color under my boat while the rain poured. Such is the life of an artist who prefers nature to art. He really should cultivate art more but feels that his love of art will take care of itself as long as it has things to feed upon." Birds were frequently the food on which my father's love of art fed. He tells of seeing a little heron climbing the trunk of a pine "using feet, wings and the point of his bill," finally reaching a branch and standing to stretch in silhouette against the sky.

> I drew it in ecstasy. It was a concentrated image which nothing could take from me. If it was not poetry, it was the image asked for by Yeats from which poetry is made. I am a painter so this morning I did two water colors of it before I got out of bed. This does not mean that I am going to be content with that one image for the rest of my life. It will generate power in me for a while, then I need another. One image succeeds another with surprising regularity on Horn Island.

The changing seasons provided an unending procession of birds. Spring meant the return of the migratory birds from South America. Year after year he witnessed and recorded the migration in words as well as images. The variety provided an "embarrassment of riches." One April evening he wrote in his log, "While I was at supper I was visited by pals from New York State, a flock of bobolinks—the males in very full dress—came and perched on rushes in the marsh. They weren't allowed to stay long. Furious redwings drove them off." A cold front moving south in March or April could precipitate a "fall out"; hundreds of tiny bright travelers found Horn Island a land fall and refuge. In its protected inner corridors, exhausted hummingbirds rested on flowering yaupon. Behind the high dunes, blooming thistles offered refreshment to painted buntings. Indigo buntings, goldfinches and scarlet tanagers lit the rain-darkened pines and warmed the artist's heart. "Nature does not like to be anticipated," he wrote, "but loves to surprise; in fact seems to justify itself to man in that way, restoring his youth to him each time, the true fountain of youth." He happily identified warblers too weary to fly and added them to the bird lists in his daily log.

The mating and nesting of resident birds demanded his full attention. It was a part of man's purpose, he wrote, "to laugh and give reason to the love affairs of the birds, himself vile, and they still happy in that first garden."

The regularity of their life cycles gave a continuity to his own life on the island. He recorded some families year after year: "I spent the morning drawing the baby heron. After lunch I successfully restored him to the nest and left him standing there staring after me. . . . I think I have drawn and painted at least three generations of his family, all of whom have nested in that tree."

There is a central curve known as the "Horseshoe" on the north beach of Horn Island. My father's regular camp was on the eastern point of this crescent. An easy crossing to the south beach lay just west of the site; directly behind it was a water lily and bullrush pond and to the east was a marsh-fringed lagoon. Each year he scouted the wetlands for the nests of bittern and gallinule, rail and wren. Grackles and herons, redwings and osprey nested in surrounding bushes and trees. When the mockingbird built her nest beside his trail, he altered his route. He searched for the camouflaged eggs of the bullbat in open clearings on the ground and for least tern and plovers' eggs on the beaches. He marveled at the courage of the kingbird's defense of its nest and grieved for the eggs lost to marauders. He regularly checked the nests he had found and noted when eggs were laid and when they hatched. On the back of a still life of eggs on a brown paper bag, he wrote, "Inexhaustible," and of a baby bittern he said, "Simply too much to come out of one small egg."

He had no compunction about borrowing the baby birds for models. He describes a green heron just hatched: "The eyes already open—little black oriental slanting pools of ink. It is dressed in an undergarment of lilac and white—lilac short feathers, white wisps more like hair than feathers."

Unfortunately he sometimes kept them too long or accidentally dropped them coming or going. He seemed genuinely sorry when one died of shock or exposure, and wrote of his feverish search when one was missing, but these mishaps didn't discourage the practice. Perhaps a clue to his persistence lies in the following: "I did a water color of the baby. I had forgotten how wonderful they were: the way the feathers come out following the construction of the

Hungry Babies c. 1965

Eggs "Inexhaustible"   c. 1955

body, the sequence of color—and the variations—" He even broke open eggs in the fervor of his study—and what artist could resist the color combination of the embryonic grackle—"bright orange with large blue globes where the eyes would be"?

Between 1960 and 1963 his interest in fledglings reached its height. He painted a complex series of baby herons, classifying them in three stages according to development. They are as bold and rhythmic as Indian dancers, as awkward and full of potential as any young rapidly growing creature. He wrote: "The grown bird is identified with a single characteristic. You can't treat the young bird that way. He is a composite thing and at the same time a marvelously complete unit—almost as much so as the egg." One of the few exhibits mounted in his lifetime, "Fledgling Birds by Walter Anderson," was held at Brooks Museum of Art in Memphis in 1964.

In summer, Horn Island's population of snakes, frogs, turtles, alligators and myriad mosquitoes, gnats and flies made it difficult to approach certain birds. The season's growth in lagoons and marshes concealed the nests and families of two of my father's favorites. One was the comical master of camouflage, his "Moses in the bull rushes"—the least bittern.

On his progresses around the edges of the lagoons or his plunges into the inhabited waters, he was always delighted when bitterns materialized: "I looked and saw a bag-like something fastened to a bullrush. I looked again and saw that it had one foot—one leg and a foot—one beak and an eye."

In his log he describes a crawl through marsh grass to peer out at the bayou's edge: "I saw a family of least bitterns fishing—two or three children and a parent—all with beaks pointed down at the water like spears. The little fish comes near enough, there is a sudden thrust and a bittern eats breakfast."

Frustrated accounts of stalking the elusive purple gallinule, another favorite, fill many pages of his summer logs:

> The marsh is full of gallinules. . . . I sketch two on the wing, then take a blind. One of them warbles to me for half an hour but refuses to come out and show herself.

> I took my walk and again was flirted with by the gallinules. I saw one and got down to crawl but when I came opposite the place it had retired into the grass. It would make noises but not come out.

Gallinules c. 1965

After one long and unsuccessful hunt, he contented himself with the song, "a gentle cluck or warble, . . . a pleasant cheerful sound accompanied by bull frogs and red wings, and bullbats and fishhawks, the wind in the palmettoes and the far away surf."

When he did finally locate a nest, a haphazard tipsy arrangement of dried rushes all grown over with magenta convolvulus, he wrote: "Just the sort of place for a pair of harlequins to nest and rear a family of little clowns to add to the gaiety of nature."

As an artist he was fascinated by the power of the gallinule's color and its effect on everything near it: "Color is simply squandered on it. It's hard for an ordinary person to know the need for a concentration of color (force). But it finds friends. Dull grey, brown logs turn to violet and gold because the purple gallinule is for an instant present."

Gulls, terns and skimmers patterned the summer skies over Horn Island or stood in flocks on the long sand spits, their heads all turned into the wind. In my father's drawings and watercolors their characteristic patterns of color and form and flight evoke quick recognition like phrases of familiar music. The flock of birds was a unit like the feather on a fledgling's wing—whole in the sum of its parts yet needing to be considered in context: "I saw . . . a pattern of terns, almost a tone, changing from white to dark in flight so that the spiral form was brought out by the values of the bodies against the sky. A mosaic of sky—of birds, of feathers: the celestial inverted cone of heaven. . . ."

Along the edges of the beach ran recurring rhythms of peeps, while low over the water the willets' black and white wings and yellow legs whirred.

My father found aesthetic pleasure in these repetitions. He enjoyed the birds' company and felt his enjoyment reciprocated: "I took a walk at dusk accompanied by a little piping plover that ran beside me. When I turned to come back he flew ahead and went that way with me. They—of all birds—but I believe a great many do—like the companionship of man."

Frequently the sandpipers welcomed his arrival to the island from Ocean Springs. After a difficult passage, he wrote that during the last pull into the

Grackle c. 1965

wind along the beach to his camp, "The little peeps running beside the boat a few feet away encouraged me."

Herons and egrets were familiar from his childhood. His affectionate names for them in the Horn Island Logs were born of this acquaintance. He called a moulty old night heron "Feathers," while the little green heron was the dramatic "Dagger with Wings." The drama of the rise of the great blue heron over the long horizontal island amazed my father each time it happened. "It is as if he were taking the sky with him," he wrote. "What is it that makes the Great Blue seem so large? It is as though part of the firmament had detached itself and through pure volition moved away. Trees become twigs: high dunes accidental disturbances of sand."

He looked at birds always as though he were seeing them for the first time. His capacity for awe and wonder seemed to intensify as he grew older and appeared most pronounced with the birds that he knew best.

His constant companions on the island were his favorite red-winged blackbirds and those most visible "lords of Horn Island," the boat-tailed grackles. His watercolors capture the grackles' strut and the comical effect of the island wind on their oversized tails, both on the beach and in the air. Their soaring silhouettes are often featured in Horn Island landscapes.

The redwings' flash and musical accompaniment were integral parts of his island, and the birds themselves were "the artist's friends in camp." He wrote: "I am surrounded by red-winged black birds. The air is full of musical motifs. They talk, scold, sing, flirt, pretending to be terribly afraid—and then asking why the rice isn't better cooked."

Red-winged Blackbirds   C. 1960

The enthusiastic welcome they regularly accorded his arrival on the island was perhaps influenced by the benefits they came to associate with his presence and even to demand: "Twice while I was under the boat waiting for the rain to stop one of the redwings came up and peeked in the doorway wanting to know why I didn't come out to feed him."

But rice was a small token to offer in exchange for company that sometimes swelled to "a symphony of birds" and almost overcame the listener with its "persistant note of joyous harmony."

He sometimes claimed that the notes they sang were a repetition of the phrase from Beethoven that he whistled as he worked. He loved redwings for their greedy appetites, their bright boldness, their cheer, their music, and their indomitable spirit: "Redwings have very little of the inheritance of chivalry but act from reason in protecting their nest no matter what the size of the bird. I've seen one set on a Peregrine Falcon."

In one watercolor of a flock of feeding redwings, an unmistakable face— bright red eyes formed by the epaulets on the wings—peers out like man's more beastly nature. Undoubtedly my father associated certain birds with human qualities because of their appearance or behavior, though he never became heavy-handed or too serious about it. His sense of humor never failed:

Dead Grebe c. 1960

Dogris c. 1955

"Coming back to camp on my walk I thought of God and a white heron flew up; and a little later when I thought of man a red-winged blackbird flushed. I also jumped a bobolink but I think my mind was blank at the time."

Many ducks winter on Horn Island. They find food and shelter in the protected inner lagoons or float in great rafts on the open water of the sound. My father looked forward to their arrival. Mallard and pintail, widgeon and teal, redhead and bufflehead offered an endless variety of winter models. He stalked them on the island as he had done in his youth hunting in the Louisiana marsh. There he had learned the French or Cajun names based on the duck's dominant feature or behavior. In his logs he calls the lesser scaup "dogris," or gray back. A merganser is a "becsie," or saw beak, and a coot a "pouldeau," or water chicken.

From bullrush or palmetto blinds, he drew ducks in groups on the water—coots advancing in battalions, iridescent mallards gliding above their reflections, dozing redheads like rounded bricks patterning the waves and dabblers upended, their heads visible through the clear cold water. He captured their whirring flight patterns against the sky, their comical landings—all feet—and their awkward waddle on the sand. Grebe, loon, cormorant, scoter, old squaw or ringneck duck—he found characteristic attributes beyond color and form that distinguished each from the other.

Occasionally he found an injured bird and kept it in his camp, nursing and feeding it and using it as a model. The opportunities for closer and more intimate knowledge of the individual bird resulted in watercolors that seem like character studies or portraits. He describes Simy, a lesser scaup with a broken wing: "He is incredibly effective with yellow eye, purple shining head, black and white zig-zag patterns on his back. He is so small and compact—reminds me of the Greek conception of soul—nature and intelligence in one."

With "captive" models he could also explore new methods and techniques: "Yesterday I did Simy in a composition twice. He composes very well together!" He found that he could record the movement of his subject in what he called "continuous composition."

He also did careful drawings and watercolors of dead birds that he found on the beach. Of these he said, "I painted well but with out joy," and that he "did honor to their remains." They afforded him new insight. He noted that the patterns in the feathers varied subtly to strengthen or accentuate each change in the form they covered as do the patterns of bark on a tree. He wrote: "It is fascinating to analyze thru distinct sequences of color in a brown and grey feathered thing."

To my father, the island, its plants and creatures constituted a whole. There were layers or levels of life on the island each complete in itself but each a part of the unity of the whole. His understanding of this interconnectedness extended beyond the physical world to the spiritual. He wrote:

The day before I had seen a cormorant on the top branch of a tall dead tree. It seemed so much part of the same organism when it flew it gave the effect of generation by fusion or separation.

The tree's exhausted life had entered the bird. About the foot of the tree are dunes and flat ponds emphasizing the horizontal flatness from which life has sprung both into the tree and from the tree to the bird.

In *The Power of Myth*, Joseph Campbell says that man hungers not so much for meaning in life but for "the experience of being alive so that our life experiences on the purely physical plane will have resonances within our own innermost being and reality, so that we actually feel the rapture of being alive." In his response to birds, my father experienced that rapture and confirmation of his own reality.

My father believed that man's salvation lay in his awareness of the natural world and his recognition of his own role in it. He wrote: "The bird flies and in that . . . fraction of a second man and bird are real. . . . He is the only man and that is the only bird and every feather, every mark, every part of the pattern of its feathers is real and he, man, exists and he is almost as wonderful as the thing he sees."

Boat-tailed Grackle   c. 1955

Birds

Nature does not like to be anticipated—it too often means death, I suppose—but loves to surprise, in fact seems to justify itself to man in that way, restoring his youth to him each time— the true fountain of youth.

Hooded Warblers c. 1960

3

4

Water Lily Lagoon   c. 1960

Tree Swallows   c. 1960

Nuthatch and Downy Woodpecker   c. 1935

Hummingbirds and Thistles   C. 1960

Painted Bunting and Prothonotory Warbler on Thistle
C. 1960

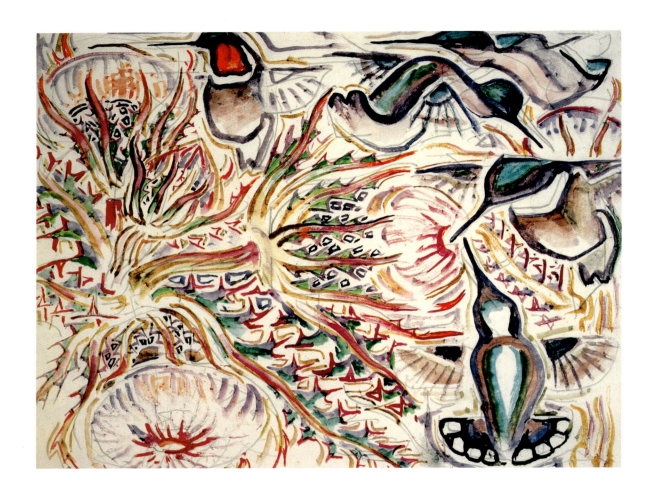

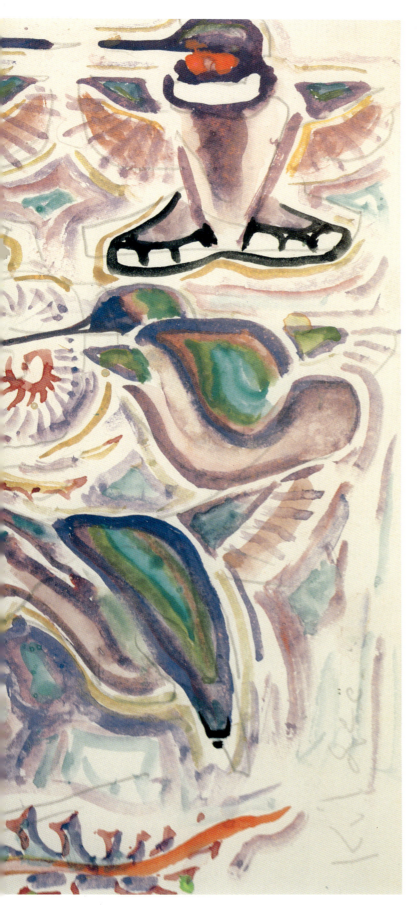

Hummingbirds in Thistles   C. 1955

Redstarts and Mangroves   C. 1955

Titmouse Plate Design   C. 1960

Hooded Warbler in the Azaleas   C. 1960

Catbirds   C. 1955

Parrots   c. 1965

Essence of Spring   c. 1955

19

Blue Jays   C. 1960

Blue Jay   C. 1935

Blue Jay Table   C. 1955

20

Birds in Spring Plate Design   c. 1960

Scarlet Tanagers   c. 1955

23

There is no precedent. All things exist in themselves, have an integrity of their own—the wind, the grass, and the little animals that move through the grass . . . .

Little Bittern, c. 1960

26

Reflections in a Bull Rush Pond   C. 1960

Nesting Green Herons   C. 1955

Green Heron, c. 1960

Pelican Chair   C. 1950

Tree Swallows   C. 1960

Mallards C. 1955

Quail in Flight   c. 1960

Calendar Drawing   c. 1940

Jan - 31.

Dead Lesser Scaup   C. 1955

"Dogris" Lesser Scaup   C. 1955

The outpouring of life must find expression in the egg.

Roc in Egg C. 1945

Nest in the Marsh   C. 1960

Baby Herons   C. 1960

Baby Bird   c. 1960

Baby Herons   c. 1960

43

Baby Herons c. 1960

Young Heron c. 1960

Baby Herons   c. 1960

Purple Gallinules   c. 1960

47

Four Baby Herons   C. 1955

Green Heron's Flight   C. 1955

48

49

Nostalgia, whether it is born of effort or memory, is a trap and only definite knowledge is of any earthly use in defeating it.

Sparrowhawk c. 1935

Kingfisher   C. 1935

Yellow Crowned Night Heron   C. 1935

Red Bellied Woodpecker   c. 1935

Pileated Woodpecker   c. 1935

Pelicans   c. 1950

Chimney Swift   c. 1935

Meadowlark   c. 1935

57

Nighthawk   C. 1935

Frigate Birds on Chandeleur   C. 1960

Birds are holes in heaven through which man may pass.

61

Hummingbird Plate Design   C. 1960

Dead Starling   C. 1965

63

Heron   C. 1965

Bird Vase   C. 1955

Loon   C. 1950

Coots and Waves Plate   C. 1965

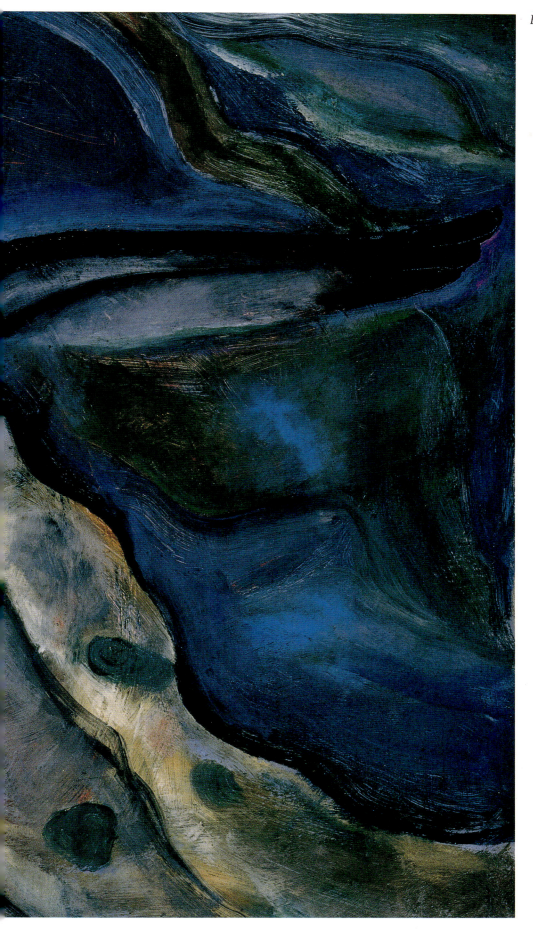

Black Skimmer c. 1930

Red Head Duck   C. 1955

Two Birds in a Tree Plate   C. 1955

Flying Blue Jays   c. 1945

Two Grackles   c. 1960

73

Hawk c. 1945

After you have lived on the island for a while, there comes a time when you realize that the pelican holds everything for you It has the song of the thrush, the form and understanding of man, the tenderness and gentleness of the dove, the mystery and dynamic quality of the night jar, and the potential qualities of all life.

In a word you lose your heart to it.

Pelicans on Chandeleur Island   c. 1955

Clouds, Pelican, and Waterspout   C. 1950

Pelican   C. 1945

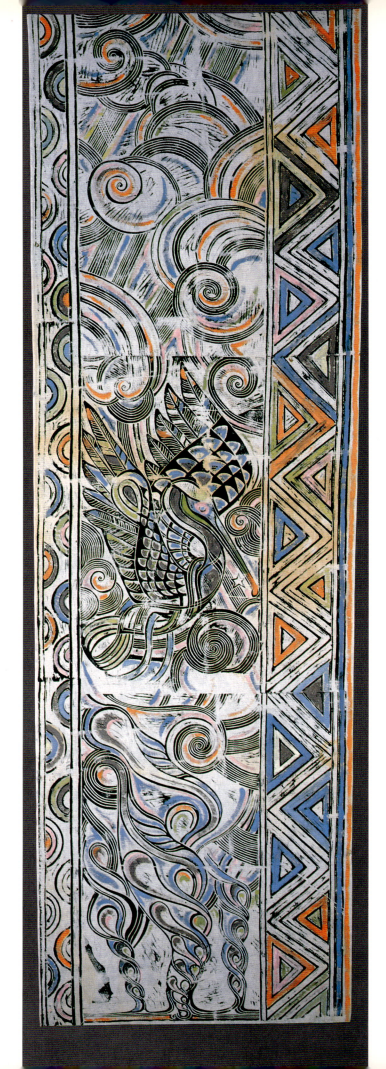

Pelican Plate   C. 1950

Dead Pelican   C. 1955

Young Pelican   c. 1955

Young Pelican   April 1950

Pelicans on Chandeleur Island   c. 1950

Pelicans on Chandeleur Island   c. 1950

Pelican Bookends   c. 1930

Pelican on Chandeleur Island   c. 1950

Pelicans on Chandeleur c. 1940

Sgraffito Bowl c. 1930

Calendar Drawing c. 1940

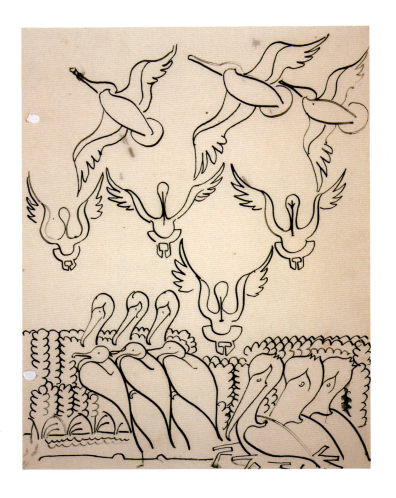

Sep. 7.

Pelicans on Chandeleur Island   C. 1945

89

Every little movement, each discovery, is part of the heavenly music and if my ears could function properly I would hear, not just the wind in the grass, the two or three different rhythms of insects, the piping of a frog, the call of a night jar, but an orderly and recognizable harmony, which might or might not have been written.

Calendar Drawing c. 1940

Oct - 15

Wrens on Grasses Plate C. 1935

Calendar Drawing C. 1940

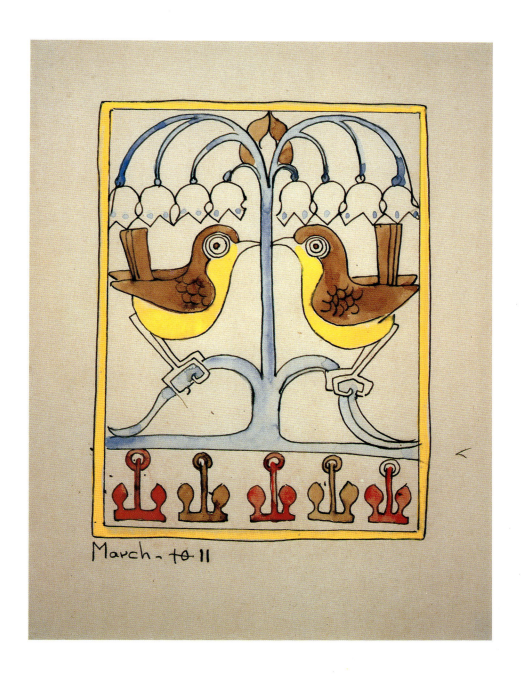

Jan-21

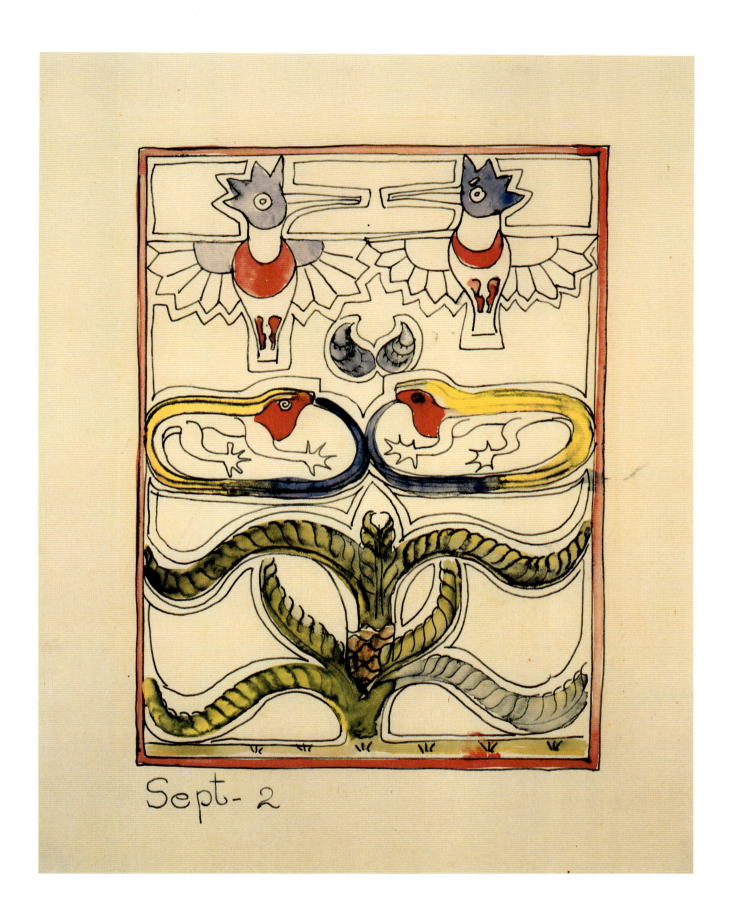

Sept - 2

Jan-10

Sept. - 1st

Calendar Drawing   c. 1940

Doves and Pecan Trees   c. 1945

The miracle of realizing that art and nature are literally one thing astonishes me each time it happens.

Redhead Ducks at Sunset   C. 1960

Hawk over Horn Island   C. 1965

The Halcyon Bird   C. 1965

Gulls and Waves   C. 1960

111

Grackles at Sunrise   c. 1965

Horn Island As a Bird   C. 1960

Birds at Sunset   C. 1960

Hawks in Sunlight c. 1960

Library of Congress Cataloging-in-Publication Data

Anderson, Walter Inglis, 1903-1965.
  Birds.

  1. Anderson, Walter Inglis, 1903-1965.
2. Birds in art. I. Title.
N6537.A48A4 1990   709'.2   90-12883
ISBN 0-87805-472-3